HOMETOWN FIRE DEPARTMENT

Buffy Silverman

Crabtree Publishing
crabtreebooks.com

I want to be a firefighter.

Firefighters put out fires and keep us safe.

They rescue people and animals.

WSTPD
FIRE-RESCUE
39
FIREFIGHTER

Firefighters ride in fire trucks.

Some fire trucks have ladders.

ENGINE
TWELVE
BRIDGEPORT
FIRE DEPT.
BFD
E-12

AIR AIR

Some fire trucks pump water through **hoses.**

Fire trucks have **sirens** and flashing lights.

Firefighters wear **masks** and special clothes.

We can call them in an **emergency**.

Firefighters are brave.

Glossary

emergency (i-MUR-jin-see): An emergency is a dangerous situation. We dial 911 in an emergency.

firefighter (FIRE-fite-ur): A firefighter is a person who is trained to put out fires.

hoses (HOHZ-ez): Hoses are long, bendy tubes that water can flow through.

masks (MASKS): Masks are worn on the face to protect or hide it.

rescue (RESS-kyoo): When you rescue someone, you save them from a dangerous situation.

sirens (SYE-ruhnz): Sirens are used as warnings. They make very loud sounds.

Index

School-to-Home Support for Caregivers and Teachers

Crabtree Seedlings books help children grow by letting them practice reading. Here are a few guiding questions to help the reader with building his or her comprehension skills. Possible answers are included.

Before Reading

- What do I think this book is about? I think this book is about firefighters. It might tell us about how firefighters stop fires.
- What do I want to learn about this topic? I want to learn more about fire trucks.

During Reading

- I wonder why... I wonder why firefighters wear masks.
- What have I learned so far? I have learned that fire trucks have ladders, hoses, sirens, and flashing lights.

After Reading

- What details did I learn about this topic? I learned that firefighters keep people safe. We can call them for help during emergencies.
- Read the book again and look for the vocabulary words. I see the word ***rescue*** on page 6 and the word ***sirens*** on page 14. The other vocabulary words are found on pages 22 and 23.

Crabtree Publishing

crabtreebooks.com 800-387-7650

Print book version produced jointly with Blue Door Education in 2022

Written by Buffy Silverman

Hardcover 978-1-4271-2954-3
Paperback 978-1-4271-2964-2

Printed in Canada/012025/CP20250108

Library and Archives Canada Cataloguing in Publication
Title: Hometown fire department / Buffy Silverman.
Names: Silverman, Buffy, author.
Description: Series statement: In my community | "A Crabtree seedlings book". | Includes index.
Identifiers: Canadiana 20200388096 | ISBN 9781427129543 (hardcover) | ISBN 9781427129642 (softcover)
Subjects: LCSH: Fire fighters—Juvenile literature. | LCSH: Fire departments—Juvenile literature.
Classification: LCC TH9148 .S55 2021 | DDC j363.37/8—dc23

Published in Canada
Crabtree Publishing
616 Welland Avenue
St. Catharines, Ontario
L2M 5V6

Published in the United States
Crabtree Publishing
347 Fifth Avenue
Suite 1402-145
New York, NY 10016

Photo credits: Cover photo © SanchaiRat, logo © AWesleyFloyd; page 2-3 © Littlekidmoment; page 4-5 © worradirek; page 6 © Utekhina Anna, 6-7 © Lilac Mountain; page 8-9 © mikeledray; page 10-11 © Keith Muratori; page 12-13 © Mike Brake; page 14-15 © welcomia; page 16-17 © zulkamalober; page 18-19 © yotanan chankheaw; phone illustration © alexmillos; page 20-21 © Monkey Business Images. All photos from Shutterstock.com

Library of Congress Cataloging-in-Publication Data
Names: Silverman, Buffy, author.
Title: Hometown fire department / Buffy Silverman.
Description: New York, NY : Crabtree Publishing Company, [2021] | Series: In my community : a Crabtree seedlings book | Includes index.
Identifiers: LCCN 2020050753 | ISBN 9781427129543 (hardcover) | ISBN 9781427129642 (paperback)
Subjects: LCSH: Fire extinction--Juvenile literature.
Classification: LCC TH9148 .S535 2021 | DDC 363.37/8--dc23
LC record available at https://lccn.loc.gov/2020050753